Francis H. Loring

New River Mines

Where they are, what they are

Francis H. Loring

New River Mines
Where they are, what they are

ISBN/EAN: 9783337240608

Printed in Europe, USA, Canada, Australia, Japan

Cover: Foto ©berggeist007 / pixelio.de

More available books at **www.hansebooks.com**

WHERE THEY ARE;

WHAT THEY ARE.

HOW TO GET THERE

BY

MOST ACCESSIBLE ROUTES.

Topographical and Outline Maps, showing relative posi-
tion of New River City and Surrounding Country,
and Towns seventy-five miles distant,
Mining Districts in Shasta, Trinity
and Siskiyou Counties.

SAN FRANCISCO:
BACON & COMPANY, BOOK AND JOB PRINTERS,
Corner Clay and Sansome Streets.
1885.

New River Mines,

TRINITY COUNTY, CALIFORNIA.

New River City.

The object of this little work is to treat of the New River Mines in a manner to do justice to all concerned. The present status of this place is as yet somewhat anomalous, being situated in Trinity County, principally prospected, so far, by Humboldt County men and capital, and by natural location belonging to Siskiyou County. To explain the statements above made by facts and figures will be our object, and at the same time present the claims of New River Mining District to a place among the great gold-producing centers of the immediate future. In order to give the reader an accurate idea of the location of these mines, and a chance to judge of the claim made that these mines are the nucleus of a richer mining district within a given circumference, radiating from this as a center,

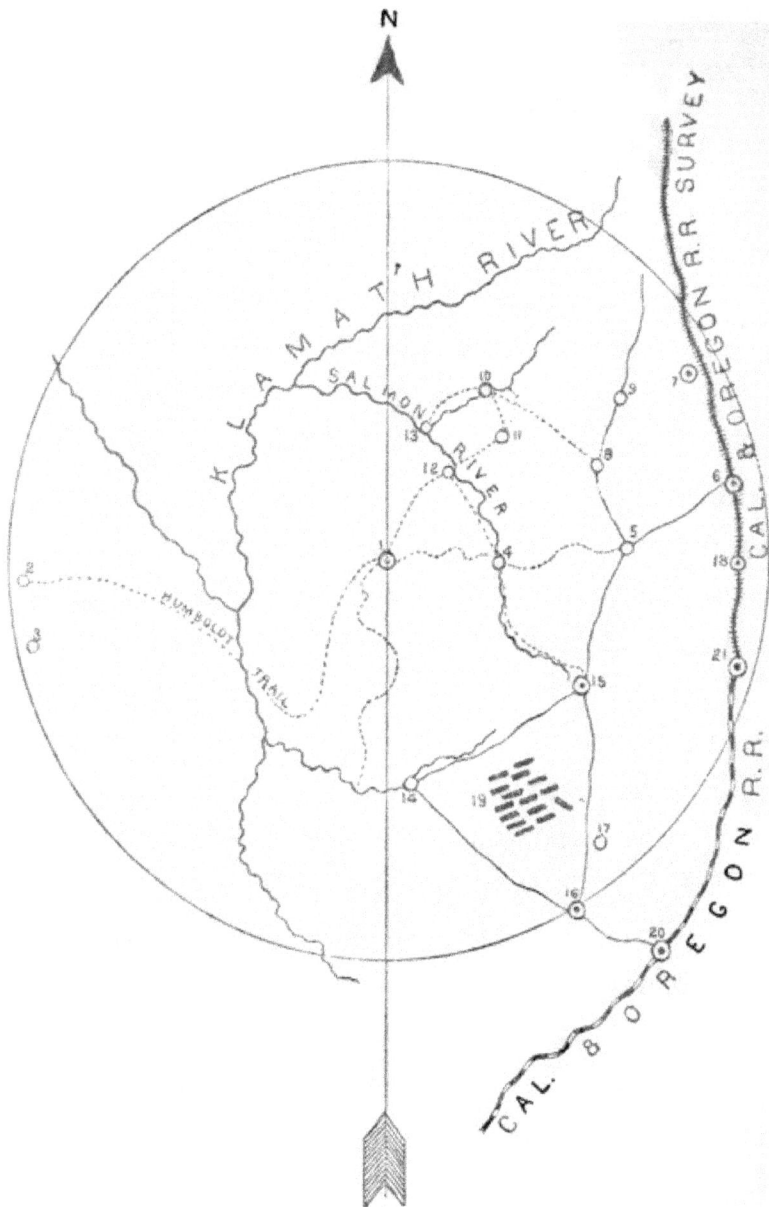

MAP No. 2.

1. New River City.	12. Yocumville.
2. Arcata.	13. Forks of Salmon.
3. Eureka.	14. Weaverville.
4. Cecilville.	15. Trinity Center.
5. Callahan Ranch.	16. Tower House.
6. Edsons, Shasta Valley.	17. French Gulch.
7. Yreka.	18. Lewiston.
8. Etna.	19. Deadwood Mines.
9. Fort Jones.	20. Redding.
10. Sawyer's Bar.	21. Delta.
11. Black Bear.	

than can be found in any other circle of like size on the Pacific Coast at the present time, we present Map No. 1 in justification of the assertion. The accompanying illustration represents a circle one hundred and fifty miles in diameter, drawn from this city as a center. It embraces the famous Yreka Flats, the Klamath River on the north, the Trinity and its tributaries, from its mouth, around by Weaverville with its wealth of gravel mines, across and including Deadwood, French Gulch, Trinity Centre ; circling north, all of Scott and Salmon Rivers and their tributaries are included. A careful revision of the mines included in this circle will show the richest, best, and most constantly paying placers and quartz mines of the State, both in the past and present. As will be seen by consulting the map, New River City is in the center of a country that has been comparatively unknown within twenty-five miles in either direction. New River itself is a tributary of Trinity River ; rising on the west side of the mountains over the divide from Salmon River and its tributaries, it flows—including branches and tributaries—southwest a distance of about forty miles, joining its waters with the Trinity about 50 miles below Hay Fork, or 35 miles below Weaverville. Ascending New River twenty miles from its mouth we reach Francis, at the point of junction of the north fork

and east fork of New River. It is the Post Office for
miners at New River City, eighteen miles distant. It
is of the north fork that we at present write. It has
been a matter of fact since the early '50's that the
gravel mines of New River and its tributaries have
been of the richest description. The men who braved
the height of the mountains, or the swiftness of the
streams, were met by the hostile Indians ; and the ex-
treme danger, as well as expense, of getting supplies,
up to 1861, had kept the miners out. About '61 so
great a rush of men was made to Pony Creek, whose
rich placers had created so much excitement, that it
was a source of mutual protection to the many drawn
together, and it was then easy to get provisions at reasonable rates. For the next three years mining was
extensively carried on in this and other portions of the
river. It was during this excitement that " Lake City "
was in full blast, and at one time claimed a population
of over four hundred within its limits. The mines being heavy to work with the then known appliances of
mining, and other mining excitements springing up,
supplemented by the Indian outbreak of 1864, Lake
City, as well as many other promising mining camps
of that period, in Trinity and what was then known as
Klamath County, became a thing of the past. Many
of the miners of those days drifted down to the river

bottom lands, found some congenial partner of the native race, selected some fruitful patch of land at the mouth of one of the numerous tributaries of the main New River or Trinity, and decided that it was easier to make a living out of the soil than to roll big boulders. Thus the present generation of miners going to the New River Mines will find the avenues of approach from the south and west sides mostly guarded by families whose parents are a mixture of races, aboriginal and Caucasian.

Since the decline of Pony Creek Mines, placer mining on other portions of the river steadily advanced, and today may be seen extensive gravel mines fitted up with the modern appliances ; and for the last three years the name of New River Mines has oft been mentioned in the mining news of the coast. Until within the last three years quartz mines were not mentioned in the New River watershed. Fifteen to thirty miles on the northwest, in the Salmon River section, quartz mining and prospecting had been going on for many years. Fortunes had been made in successful strikes, and many fortunes lost in mismanaged and unsuccessful strikes.

It was from these last causes Mr. Oliver Clement was induced to try the New River side. Mr. Clement is the first successful prospector and miner in the district, and it is due to his perseverance and success that New River City has a place today. **Mr.** Clement is a young man of genial nature, and he has always been termed a lucky miner, having made several strikes on the Salmon side when his fellew miners would pass by the same ground. His first profitable strike in quartz was in the Mountain Boomer Mine, where a vein of free gold quartz was found; and at this point, in the summer of 1884, two partners were admitted to an equal interest in the mines, and an arastra of small capacity was erected, and to this writing has paid over three thousand dollars per month, or ten thousand dollars in all.

The discoveries made from time to time, while Clement & Co. were prospecting and erecting the arastra on the Mountain Boomer, made such an impression on many Humboldt men of means and leisure, that many men were sent into the district to prospect on shares, and many others went on their own account. The location of the district, as far as now known, extends from Slide Creek, the Mountain Boomer being on the east side of Slide Creek, northwest to the head waters of Eagle Creek, a distance of about six miles, and from the divide, between the Salmon River and New River tributaries, extends west about six miles, thus embrac ing a portion of Trinity County adjoining the County

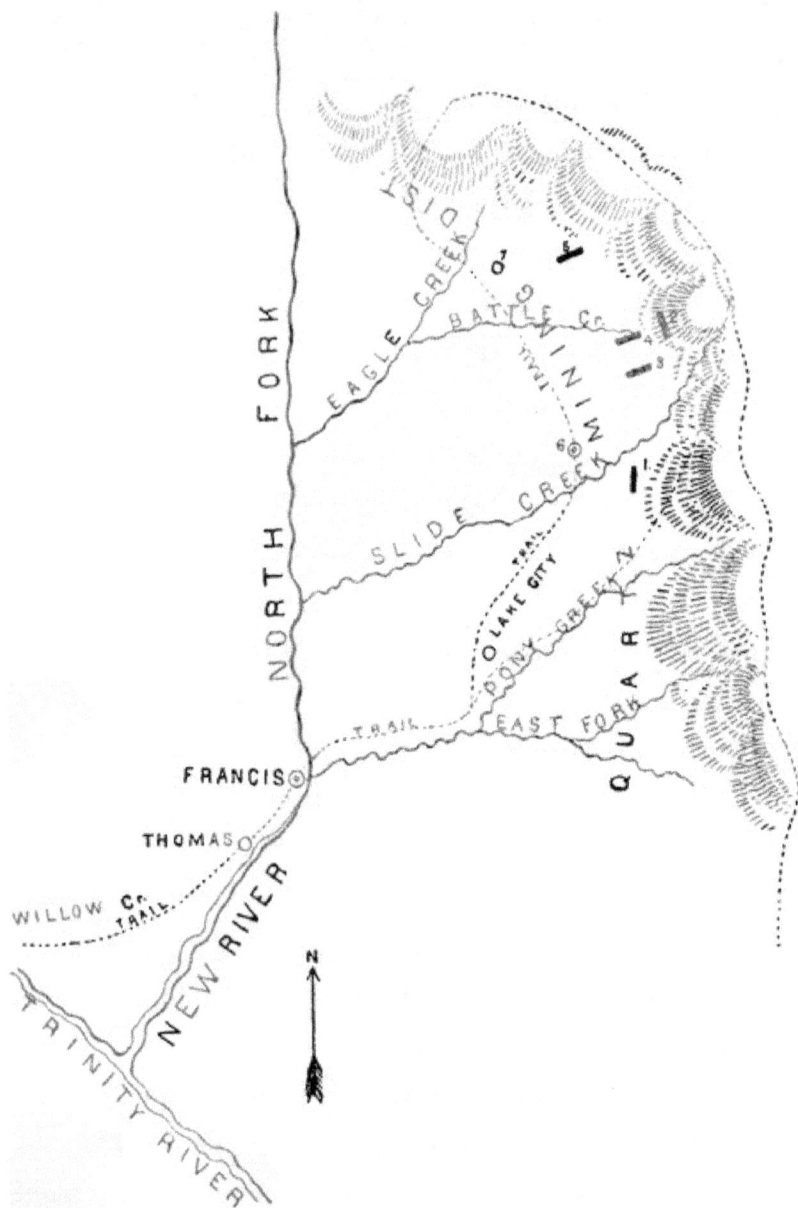

NORTH FORK

DIST

EAGLE CREEK

BATTLE CREEK

MINING

SLIDE CREEK

TRAIL

LAKE CITY

PONY CREEK

QUARK N.

EAST FORK

TRAIL

FRANCIS

THOMAS

WILLOW Co.
TRAIL

NEW RIVER

TRINITY RIVER

N

MAP No. 3.

1. Mountain Boomer Mine.
2. Mary Blaine Mine.
3. Hardtack Mine.
4. Ridgeway Mine.
5. Bolles Mine.
6. New River City.
7. White Rock City.

of Siskiyou. By referring to the map, it will be seen that Pony, Slide, Eagle and Battle Creeks are on the east side of the north fork of New River, and all of them extend east, up to the divide separating them from Salmon River, and which forms the boundary line between Siskiyou and Trinity Counties. On the Salmon side, parallel creeks are Cecil, St. Clair, Plummer, Methodist and Know Nothing Creeks, all of which have proved very rich in placer mines.

Having finally located the New River Quartz district between Pony Creek on the southeast, Eagle Creek on the northwest, Salmon River Divide on the northeast, and north fork of New River on the southwest, we find about thirty-six square miles of territory. New River City is a fine location on a flat near the principal mines on Slide Creek, about sixteen miles from Francis P. O., situated at junction of north fork and east fork of New River. A town site has been laid out by U. S. Surveyor, and patent applied for, and lots are now for disposal by the New River Mining and Development Co. In the territory above mentioned, over one hundred locations have been made during the past season, all of which claim to have assays ranging from two hundred dollars per ton up.

From the recent discovery of most of the ledges,

very little development has been made, and lateness of the season has prevented many improvements already projected. Fifty buildings of all descriptions have been erected for mining and residence purposes. White Rock is the name of the first location made for town purposes, and boasts of two stores, hotels and saloons. White Rock is situated about two miles northwest of New River City, in close proximity to several very promising quartz locations, among which are the Bolles & Co. ledge and mill, the Dean & Co. ledge, and others.

Among the improvements already made is a stamp mill and saw mill run by steam, for the use of the Mary Blaine and White Elephant locations. The Hard Tack Mines, owned by Clement & Ladd, have a fine new arastra, run by a large, overshot wheel, and with room on shaft for belt for running another arastra, or stamp mill, for purpose of crushing ore, of which they now have in sight one hundred tons that will no doubt pay two hundred and fifty dollars per ton. At the mine they have entrance covered, and a large building for work and for holding ore erected. On same ledge adjoining is the Ridgeway location, where full preparation is made for taking ore out early in spring, to be worked in their eight-stamp mill now on the way.

Numerous saw mills are on the way or projected. The country is well timbered with pine, fir and cedar timber, good to saw or split. Water is abundant for all necessary purposes. Most of the lumber used so far is whip-sawed, at six dollars per hundred feet; shakes split out at eight dollars per thousand; plenty of small timber for building purposes. Elevation ranges from 3,500 feet at New River City to 5,400 feet at the Mary Blaine Mine, which is the highest elevation of any of the mines, and nearly at the top of Salmon River divide. Up to Dec. 1st no frost has appeared in the central portions of the district, although at Mary Blaine and low down on the river frost has made an appearance. The climate is delightful—air dry and bracing, tempered by the fragrant spruce and pines. The water is pure, and the streams and springs are never dry. Snow is said to lie about four feet deep in winter, and melts off quite early in April, so as to admit of prospecting. Still, on the divides on all sides the snow is often impassable until May. About one hundred men are here for the winter, over one hundred having left because of the inability of parties to get in enough supplies. The knowledge of this fact has done more to repress information being given to the public than any other cause.

It was known early in September that it was impossible to get enough supplies into camp, before heavy snows, to last those through the winter that would like to remain. All the pack trains of Humboldt, Trinity, and Siskiyou Counties were engaged ahead, with their regular packing; and, although offered from one to two hundred per cent. advance on usual price of packing to this point, only a limited amount has been obtained at these figures. Therefore those acquainted with the situation have discouraged all new comers as far as they could, as all such helped to diminish the supplies of food and forage on hand and to come. It will not be advisable for parties to come here before the 1st of June and depend upon finding supplies, as they cannot be packed in here much before that time, and the camp will be destitute of supplies by early spring.

As will be seen by the following price list, supplies of some staples are short: Flour, per hundred pounds, eight and nine dollars; hay, feed of six pounds, fifty cents; oats, feed of six pounds, fifty cents; bacon, ham, and lard, per pound, twenty-three cents; potatoes, apples, and onions, per pound, eight cents; butter, per pound, thirty-five cents; cheese, per pound, twenty-five cents; fresh beef, per pound, ten cents; fresh mutton, per pound, ten cents; board and lodging, per day, one dollar and fifty cents. Mining companies have been paying twenty-five dollars per month, per man, for board for their entire number of men.

The price of living at this point is as cheap, all things considered from a natural basis, as at most of our older towns or mining camps.

The scarcity of transportation this fall has doubled the usual cost of hay, grain, flour and vegetables, but upon the opening of the trails in the spring, competition will be the order of the day among the packers, as their busy season does not commence until fall, and trains will be comparatively idle through spring and summer.

So far, our acquaintances and connections have been mostly from Humboldt County. Our beef, pork, butter, beans and potatoes come from Humboldt County and Lower Trinity River, making a much needed outlet for these places, heretofore remote from a paying market.

The cost of freight from San Francisco to Eureka, or Arcata, is from two dollars and fifty cents to five dollars per ton, and time from three to six days. From Eureka or Arcata here the time is for freight about eight days, or fourteen days for round trip of pack train. Lowest cost of freight in summer is four cents per pound. This fall, the price has been five to five and one half cents per pound. Our Trinity County source of supplies so far has been from Willow Creek and a number of farms on Lower Trinity, who have furnished us some vegetables, fruit, beef and pork ; and although we are comparatively near to Weaverville, the county seat, no available route for supplies has so far been used.

Our next base of supplies is Scott Valley, in Siskiyou County ; Callahan Ranch, distant forty-three miles, and Etna, distant fifty-two miles, being the two available points of supplies. At these points we get hay, grain, fruit, vegetables, flour, beef and pork, and all farm supplies at farm cost.

The basis price in summer for transportation from these points to this place is two cents per pound, and time for round trip is seven days, thus saving in cost and time one-half over the route to Humboldt Bay.

With this camp, as it is everywhere, the almighty dollar is to decide all questions of expense, profit, and routes of ingress and egress to the future Virginia City of California. The place is now like the helpless infant, and its steps are directed by its fosters ; and as it gets able to assert its own force, it will take the best and most natural one for its sustenance. It is hard to find a location with more natural barriers to access to or from its sources of supplies. East, west, north and south are high mountains and swift streams, and expensive means of transportation ; but the coming summer will teach us the most available route, by stimulating the authorities of the surrounding counties to efforts to secure some of the business that is sure to come.

So far, nothing has been said of the quality of the ore found in this district, the nature of the formation of the mineral-bearing lodes, or the extent of the deposits. It is not the intention of this writing to state something that is not known, and in this new camp no new developments have yet been made that will warrant any specific analysis of the formation to be given. In the first mine worked we find free gold, easily saved after being ground out in an arastra. Quartz is decomposed in a granite and limestone formation.

In other mines showing free gold and assaying high we find the vein of decomposed ore in porphyry and limestone formation ; other places serpentine and limestone on either side of rich ore, and another location is ribbon quartz in slate and serpentine. Some ore with silver indications has been found. Sulphuret ore has as yet not been found in extensive deposits. One location has been made on ore rich in copper.

Out of the one hundred or more locations already made, we find at least six mines in which sufficient developments have been made to entitle the owners to feel they have a fortune in sight. One mine upon which a five stamp mill is erected is selling at the rate of fifty thousand dollars, and another one which has one hundred tons of rich ore on the dump, and no very expensive improvements upon it, would be considered cheap at sixty thousand dollars.

This is emphatically a poor miners' camp, as in all the ledges found the gold is free from the surface down, and can be made to pay wages with a hand mortar; but so many of the men here are backed by capital, it is not necessary for them to wait the tedious process of a hand mortar; but as soon as they can show a reasonable prospect, ample means are at hand to help them to develope their find, wages paid.

The rate of wages paid is in proportion to the kind of work done. A few cooks are hired at one dollar per day. The highest wages paid for drifters to date is two dollars per day. Carpenters for rough work get the same price. Mechanics for building arastras and erecting water wheels and stamp mills, from three to four dollars per day.

As in all new camps, most of the workmen are on the prospect, and as soon as they get through with a job of work the money received for it is spent prospecting; and if nothing is found, they are ready for more work by which to make another raise.

The Chinaman, to date, has not made his appearance in the district, and it is the determination of the white settlers and prospectors to keep John out. When asked the reason for it, one of the owners of a rich dump replied: " You see that pile of ore there? Yes ; well, if the Chinese were here they would have a fortune in rich rock packed off before spring."

Looking at the question in this light, the answer was a good one and correct. Not long after, meeting an old Chinaman, he was asked if he knew where New River was. " Oh, yes, me sabe New River. Twenty years ago plenty Chinamen work in New River, eighty-five miles from Weaver. Me pack two mules, sell cargo to Chinamen. Injun man he killum my two Chinamen ; me put one down in ground, pack one man, go back Weaver. Loseum cargo—me lose two, three thousand dollars. Allee Chinamen lun away. In July, Augusty, Injun man he kill my two Chinamen, and at Cox's Bar he kill four Chinamen. Me sabe big Injun ; he got two squaw. White man Jim, he kill him, and pretty soon white man killed Jim. Me sabe —he got Injun woman, plenty of little boy, little girl. One place three man work up mountains ; not much water ; very good pay, I think. Oh, yes, me heap sabe New River. Too much Injun man, no good." A good illustration of the reputation the district has enjoyed for the past twenty years.

While it is not necessary to inform an old miner what he will need to possess himself of, after his arrival here, to make himself independent in the way of board and lodging, and to prosecute his work of developing the mineral wealth of the camp, for the benefit of those who may make this their initiatory trip to the mines, we will suggest as necessary articles in the bill of fare and cooking utensils : One sack flour, one frying-pan, tin plates, tin cup, sheet iron pan for wash basin, mixing bread in, and prospecting or panning out rock in, one eight-quart iron camp kettle, knife, fork, and spoons, side of bacon, beans, salt, pepper, tea, sugar, coffee, baking-powder, dried fruit, matches, axe, pick, and shovel—these all to cost about twenty-five dollars. A tree, brush house, or canvas tent are the most available shelters for frequent change. With the above outfit a man can get along from three to fifteen days, according to distance from his base of supplies ; and if he finds something to justify him to add more to his stock of provisions, by reason of more time necessary to develop it, cheese, condensed milk, butter, vegetables, and fresh meat are among the obtainable articles in a miners' store.

How to get there. There are numerous routes by which this place can be reached. The Humboldt and lower Trinity County travel will come by way of Mad

River, Hoopa Valley and Trinity River to Thomas Place, at Francis; thence by good trail to New River City. Time from Arcata by saddle horse or mule is four days, and good stopping places are along the route at convenient distances for refreshment and entertainment of man and beast.

Persons travelling through this section can figure on an expense of about three and one-half dollars per day for self and animal. The trip is very enjoyable to one used to saddle exercise, as the rider passes fine farms and orchards, crosses swift streams, and surmounts steep mountain trails and high and narrow grades around precipitous points of rock.

The Humboldt people are making real effort to control the trade of this section, by liberal gifts of money for extending a wagon road, and improving trails in this direction. For travel outside of the Humboldt route, we will name Number Two, or direct route from San Francisco or Sacramento, take the Cal. and O. R. R. to Redding; thence by stage forty-eight miles to Weaverville, the county seat of Trinity County. Weaverville is at present the center of quite an extensive quartz industry.

The gravel mines of Weaver Basin have long been of the best and regular paying class. The formation of its surrounding hills and from discoveries already made entitle it to be, as it always has been, one of the richest and prettiest mountain towns in the State.

The quartz interest of Weaverville has been steadily growing for the last five years. Its citizens have been enterprising prospectors, and many of them have been rewarded by rich finds in the Deadwood, North and East Forks of Trinity Mines.

The newcomer is warmly welcomed, and he is made to feel that the residents of Weaverville are as hospitable as they are enterprising. Time from Redding to Weaverville is about ten hours. Fare from San Francisco to Weaverville, $16.50.

From Weaverville, by inquiry of the livery stable keeper, a horse or mule can be obtained for the trip to New River. The hire will be about two dollars per day for the animal for the trip. By wagon road to North Fork, where is a good hotel, the distance is fifteen miles; from this point to Rattlesnake by good trail it is twenty miles; thence by trail to Grizzly eight miles; from Grizzly to head of Pony and Slide Creeks, or the New River Mining District, it is about sixteen miles—a total of about sixty miles from Weaverville.

No trail is as yet constructed over this last portion of the route, but by early summer no doubt a good trail will be in condition for travel, and the distance reduced by several miles. Travel has already passed over the whole route above mentioned to New River, and to one acquainted, or with a guide, no trouble is found in making the trip, and arriving at New River City the second day from Weaverville.

While making this trip from Weaver the traveler will pass through one of the most important quartz sections·of Trinity County, known as the Enterprise District. It is a link in the long chain of quartz croppings and mines extending from Salmon River Mines on the north, through New River, to Deadwood, French Gulch, and other Shasta County Mines on the south, a distance in a direct line from north to south of about one hundred miles.

The East Fork of the North Fork of Trinity already show good mines, and good prospects on Grizzly are found, and Rattlesnake will be one of the next important quartz districts. The next important route from Redding to New River is by stage to Trinity Center, distance fifty-one miles, thence by way of Coffee Creek trail to Summerville, thirty-two miles distant, thence down the south fork of Salmon River twelve miles to Cecilville, thence by trail from Cecilville seventeen miles to New River; total distance from Trinity Center to New River, sixty-one miles.

By the Trinity Center route we pass well developed and paying mines at French Gulch, and gravel mines at Trinity Center, also on Coffee Creek; and on Coffee Creek quartz discoveries have been made the past year that indicate paying ledges, arrangements having been made for their development in the spring.

In the vicinity of Summerville we find a quartz ledge already worked on a small scale, and one of the best fields for prospectors for quartz. At this point is an extensive gravel mine, the situation of which for advantageous working is perfect. Plenty of fall for and exclusive water rights, a large deposit of paying gravel, combined with favorable facilities for lumber and procuring supplies, make this a model mine, and one of sure remuneration for all work done. The location is among the grand snowy peaks of the Salmon range, and the privilege of being in constant view of such gems of nature is a recompense for isolation from the outside world.

From this mine at the head of the South Fork of the Salmon to Cecilville is ten miles, and there are many good indications of quartz; and one location with flattering prospects is now being developed two miles from Cecilville. There are parties here who have other ledges to prospect in the spring. Many old placer miners in this section (Salmon) have found gold-bearing ledges near at hand, in years gone by; but as they were engaged in placer mining, and not knowing anything about quartz mining, never made any attempt to prospect them.

Since the New River discoveries many old finds are being uncovered, and a new lease of life seems to have taken hold of the old miners in the prospect of finding something rich in quartz in the spring.

Another practicable route to the New River Mines is by way of Redding terminus and stage eighty miles to Callahan Ranch; thence by wagon road five miles, and trail five miles to the summit of the Salmon divide; thence down the East Fork of South Fork of the Salmon River eighteen miles to Cecilville; thence seventeen miles to New River—a distance from Callahan Ranch of forty-five miles by a good trail, the portion to Cecilville having been traveled over twenty years. Time in summer from Redding to Callahan's is sixteen hours, and the distance to New River has been traveled on horseback in thirteen hours, or one day's ride.

This route enables parties wishing to communicate with San Francisco to go from here to Callahan's in one day, and send by telegraph and receive an answer the same day, there being a telegraph office at Callahan's; also Wells, Fargo & Co.'s Express and Post Office. The trip by this route can be made from San Francisco in three days. When the California and

Oregon Railroad is extended through from Delta, Edson's Station, fifty-five miles north of Delta, will be the distributing point for the southern end of Scott Valley, and will be seventy-three miles distant from New River. The route, then, from this place will be thirteen hours by horseback to Callahan Ranch; thence five hours to Edson's, and by rail to San Francisco, making it possible to travel the entire distance to San Francisco in two days and one night.

Ten miles from Callahan Ranch, southwest, is the summit of the Salmon range between Salmon River and Scott Valley. Seven thousand feet high at this point, we leave the trail to New River, and turning to the right we follow the divide, and in a distance of ten miles we reach the Uncle Sam Quartz Mill and Mine. A short distance further, and almost adjoining, is the Klamath Mine and Mill. About four miles further is the Black Bear Mine and Mill. Five miles northwest of Black Bear Mine is the great mining town of Sawyer's Bar.

Returning to the summit, we travel on down the east fork of the Salmon to Six Mile Creek, and on the route down portions of the side hills are covered with float quartz. Eleven miles from the summit we reach two extensive gravel mines fitted up with little giants and large pipe, and worked by the hydraulic process.

The East Fork forms a junction with the South Fork three miles from these mines, and three miles further, on the east bank of the South Fork, is Cecilville, distant twenty-eight miles from Callahan Ranch. At Cecilville is a Post Office, store, and hotel. A mail comes once a week from Trinity Centre; also from Yocumville, fourteen miles further down the river.

This is our stopping place for the first night from Callahan's, if you are too tired to ride through. From Cecilville to New River, distant seventeen miles, the time is six hours. We ride down the east side of the south fork of Salmon River two miles, and cross to the trail, which winds up the face of a steep hill; and when on top of the ridge, some of the finest views of Salmon River scenery are to be had. This ridge lies between St. Clair and Plummer Creeks, and is about five miles long, and so narrow one can look down into either creek while riding along.

This was a favorite route for the Indian travel in their trips from Salmon to Trinity and New River; the high trail enabled them to see up and down the Salmon for miles, and the unsuspecting enemy was exposed to their strategy and cunning. In this country we are indebted in a great measure to the knowledge of the Indians for the routes of travel through these mountains —for the easiest trails that we at present travel, as the

white man has succeeded the red man as a traveller
of the steep and rocky.

A striking feature in the formation of this long ridge
is a large quartz lode that is visible its entire length,
and in places where parties have prospected they
claim to have found free gold. If such is the case, it
will be a profitable ledge to work, as from the narrow-
ness of the ridge it seems like a ledge set up on edge
entirely out of ground. Ascending this ridge, we reach
the first summit, which is called Limestone Peaks, a singu-
lar formation of marble limestone, rearing its long line
of snowy summits to a height of over six thousand feet;
and some of them are partially covered by evergreen
shrubbery. A person is reminded of the many mounds
and marble shafts of Greenwood Cemetery or Lone
Mountain. Over the summit we descend to Plummer
Creek, " and find the first drop of water since we left
Salmon River, ten miles," and then make the ascent to
summit of New River water-shed, about six thousand
feet high at this place. A short distance over this divide we pass some wild and rocky points, and come
into view of the first proof of the New River Mines,
the works and dwelling belonging to the Mary Blaine
Mine. Another route from Scott Valley is by way of
Etna, " twelve miles northwest of Callahan Ranch."
To Sawyer's Bar it is twenty-five miles; thence five miles

to Black Bear ; thence down Black Bear Creek to Yocumville, on south fork of Salmon, seven miles ; thence by Methodist Creek up to divide at head of Eagle Creek, ten miles; thence to New River City, six miles, a distance of fifty-three miles. By this route the traveller can see some of the best gravel mines of the north fork of Salmon, also pass by the famous quartz mines and mills of the Salmon : the Uncle Sam, Klamath and Black Bear, all of which are well developed, and permanent and paying mines.

The gravel mines of the Salmon are immense deposits of heavy wash gravel, and are lasting and profitable. To the lover of natural scenery, this is one of the grandest and most impressive portions of the State, and where pleasure and profit can be combined. While there are other routes than these mentioned by which this place can be reached, they are local ones, and cannot be used to advantage over the ones already mentioned.

ROUTE FROM HUMBOLDT BAY.

By Boat from Eureka to Arcata, and Cars to Mad
   River.. 24 miles

Mad River to Lupton's.............................. 14   "

Lupton's to Berry's, north side of Redwood Creek..... 4   "

Redwood to Willow Creek. 14   "

Willow Creek over Happy Camp Mountain to Thomas's
Place... 20 miles

Willow Creek to Thomas's Place, by way of Hawkins's
Bar.. 16 "

Thomas's Place to Mines........................... 25 "

Hawkins's Bar to Thomas's......................... 14 "

    Total..115 "

Time, four days ; meals en route 50 cents ; horse feed, 25 cents.

WEAVERVILLE ROUTE.

Weaverville by road to North Fork..................15 miles

North Fork by trail to Rattlesnake....................20 "

Rattlesnake to Grizzly............................... 8 "

Grizzly to New River...............................17 "

    Total..60 "

Time, two days ; meals, 50 cents ; horse feed, 50 cents for
hay, 50 cents for grain.

TRINITY CENTER ROUTE.

Trinity Center to Carville........................... 4 miles

By trail, Carville to Nash Mines, Coffee Creek........16 "

Nash Mines to Somerville, on South Fork of Salmon
River..16 "

Somerville to Cecilville............................. 8 "

Cecilville to New River.............................17 "

    Total..61 "

Time, two days ; meals, 50 cents; feed for horses, 50 cents
for hay, and 50 cents for grain.

ROUTE FROM CALLAHAN RANCH.

Via South Fork of Scott River to summit of Salmon
  Divide...10 miles
Divide to Cecilville, "by way of East Fork of Salmon
  and Brownsville"...............................18 "
Cecilville to New River.............................17 "
    Total.....................................45 "

Time, one day; meals, 50 cents; hay and grain, 50 cents per feed.

First class, Rail and Stage fare from San Francisco to Red-
  ding..$ 9.05
To Callahan Ranch............................ 10.75 $19.80
Railroad ($9.05) and Stage ($7.25) from San Francisco to
  Weaver... 16.30
Railroad ($9.05) and Stage ($7.50) from San Francisco to
  Trinity Center................................. 16.55
San Francisco by steamer to Eureka, first-class...$12.00
Eureka to Mines............................. 15.00 27.00

If people wish to travel second class by these routes, and walk from terminus of stage and steamboat travel, they can make the trip for nearly one-half the rates mentioned; but as twenty-dollar pieces are not lying on top of the ground at New River, waiting to be picked up, we would advise all who wish to make the trip to be sure and have coin enough to pay your way until you leave the camp, or find work or a paying mine. The old style of staking the *honest miner* has been out of fashion in all new camps for some time.

While this camp is yet new and undeveloped, and very little gold has been sent to mint from the quartz mines — a little less than ($10,000) ten thousand dollars having been sent— the preparations already made to work the ore in sight almost warrant us in estimating the yield from four mines, within six months, at seventy-five thousand dollars. The universal opinion of men and mining experts, who have been here, and given much time to examination of the district, is that while developments already made will not warrant them to declare an opinion favorable to the great depth and permanence of the veins, they are well pleased with the outlook for the richness and permanence of the camp, and *all* express a desire to come back in the spring, or soon as weather can be depended upon for out-door life.

For placer mining there is, no doubt, an opening in the tributaries of the north fork of New River, and the coming summer will see a revolution in the mining interest of this section. The title to all land other than the mineral portion is still vested in the U. S. Government, as we are outside of the range of the railroad grants. In fact, the whole country here for

scores of miles is in its natural state, unappropriated
and unknown. The coming tide of gold-seekers will
develop its resources, and settle the present questions
of difficulty of ingress, and means of access to the
depots of supplies.

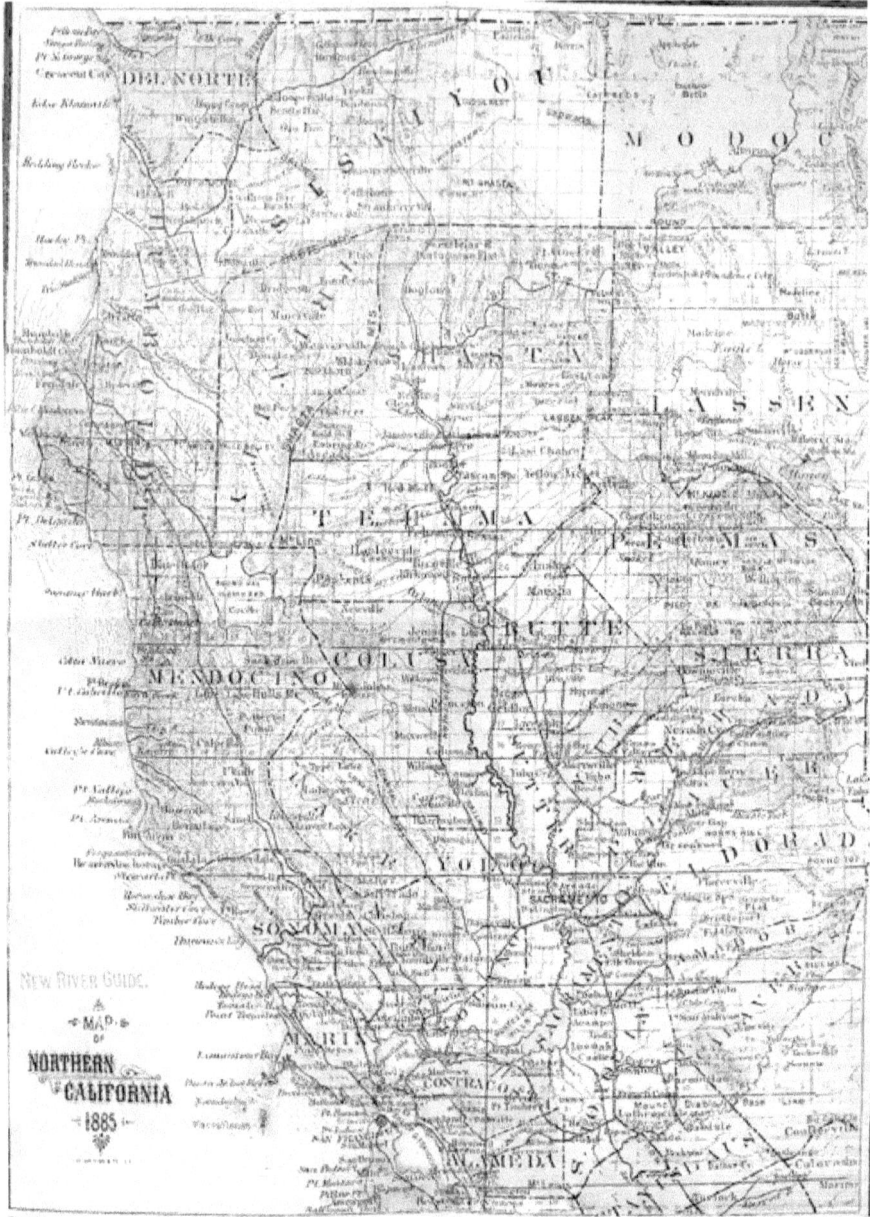

NEW RIVER GUIDE.

MAP OF NORTHERN CALIFORNIA

1885

www.ingramcontent.com/pod-product-compliance
Lightning Source LLC
Chambersburg PA
CBHW022033190326
41519CB00010B/1692